To:

_____

From:

_____

United We Stand Series

SEALs

Naval Special Warfare in Action

SF Tomajczyk

Dedicated to SEALs and SWCCs
and to all the men and women quietly involved
in Naval Special Warfare:
You keep America strong, safe and free.

Thank you for your sacrifice.

An imprint of
Turning Point Communications, LLC

Post Office Box 7070
Loudon, New Hampshire 03307
www.CalltoArmsBooks.com

Copyright © 2014 SF Tomajczyk
All rights reserved. Reproduction of the contents in any form, either in whole or in part, is strictly prohibited without prior written permission from Turning Point Communications, LLC, except for brief quotations of less than 200 words in critical articles and reviews.

To the best of our knowledge, the information in this book is true and complete and comes from unclassified, publicly available sources. All recommendations are made without any guarantee on the part of the author or publisher, who also disclaim any liability incurred in connection with the use of this data or specific details. We recognize that some words, model names and designations, for example, mentioned herein are the property of the trademark holder. We use them for identification purposes only. This is not an official publication.

Front Cover – *A SEAL barely breaks the surface of the water to do a quick "sneak and peek" of the surrounding area. On a real combat mission this would be done under the cover of darkness, but for training purposes it suffices.*

Frontispiece – *A Mk-8 MOD 1 SEAL Delivery Vehicle prepares to launch from a Dry Deck Shelter attached to a submarine. The mini-sub is used to transport the pilot, navigator and four combat-armed SEALs on a variety of missions, including reconnaissance, sensor placement, mine warfare, hydrographic survey, assault, insertion/extraction, and intelligence gathering. The SDV and its crew are considered to be a Tier I National asset, a highly specialized and vital warfare tool for the United States. It joins Delta Force, Night Stalkers and DEVGRU ("SEAL Team Six") in that role.*

Title Page – *A SEAL practices low-light close-quarters combat techniques in a Kill House.*

Table of Contents – *SEALs are the maritime component of America's special operations forces. They are uniquely trained to attack from the sea, from the air, and across the land.*

Library of Congress Control Number: 2014931332

ISBN-13: 978-0-9911198-1-3

First Edition   Printed in the USA

CONTENTS

Introduction — 6

Making of a Warrior — 12

From the Sea — 40

From the Air — 74

Across the Land — 98

Glossary — 120

"In times of war or uncertainty there is a special breed of warrior ready to answer our Nation's call. A common man with uncommon desire to succeed. Forged by adversity, he stands alongside America's finest special operations forces to serve his country, the American people, and protect their way of life."

EXCERPT FROM
THE SEAL CREED

Operating in small numbers of just two to eight men for stealth and speed, SEALs are maritime specialists who are experts at reconnoitering and clandestine attacks. They are trained to go in, do a job, and come out without anyone knowing they were there. On any given day, SEALs are deployed in more than 30 countries around the world, and nearly every night they are involved in some type of joint special operations mission.

SEALs are known for their physical strength, mental toughness, ingenuity and bravery. Rejecting comfort and glory, and operating in the most challenging and unforgiving environments known to Man, SEALs control any fear or anxiety they might have to focus on accomplishing their mission and not failing their teammates. As far as they are concerned, it's "All In" – All the time. All the way.

As a result, SEALs have earned their place on the battlefield along side America's other elite fighting forces, and in the history books.

With public attention focused on the SEALs, especially in recent years with so many incidents being handled by the SEALs, people fail to understand that these men represent

Introduction

Before deploying, SEALs can plan and rehearse a mission using SOFPREP, a software program that creates a 3D simulation of buildings, terrain, vehicles, etc. based on geospatial intelligence (photographs, maps). That way they know the area they will fight in before arriving. It is especially helpful for night operations.

only a small portion of Naval Special Warfare. There are about 10,000 personnel assigned to NSW, of which only 2,700 are active duty SEALs. The remainder represent the men *and* women behind-the-scenes who support SEALs on their missions, including 750 Special Warfare Combatant-craft Crewmen (SWCC) and 4,400 support personnel ("enablers"). The latter includes intelligence analysts, aerial drone pilots, linguists, cultural experts, divemasters, meteorologists, information systems technicians, parachute riggers, corpsmen, and operations specialists. SEALs are also joined in the field by construction and logistics personnel from the NSW Combat Services Support Team who build facilities (camps, command posts, mock buildings for training purposes) and coordinate the movement of "beans, bullets and band-aids."

This expansive network of dedicated people is truly what ensures the SEALs' success in combat.

America has more than a dozen special operations forces, ranging from Army's "Green Berets" and "Night Stalkers" to the Air Force's "Combat Controllers" and "PJs." Each unit brings its own unique fighting capabilities to the mission-planning table. The Green Berets, for example, are known for training and leading unconventional warfare forces (or a clandestine gorilla force) in an occupied country. By contrast, members of the 160th Special Operations Aviation Regiment ("Night Stalkers") fly highly-specialized helicopters at night using night-vision goggles to insert or extract commando forces.

These elite units are deployed on combat missions around the world by the United States Special Operations Command (SOCOM) in Tampa, FL, which reports directly to the President and the Secretary of Defense. The units are represented within SOCOM by a liaison command. For the Navy, it's the Naval Special Warfare Command (often referred to as WARCOM).

Headquartered at Naval Amphibious Base Coronado just across the bay from San Diego, CA, the Naval Special Warfare Command recruits, trains and prepares SEALs, SWCCs and mission-support personnel for war. The Naval Special Warfare Center (NSWC) plays a vital role in this endeavor since it oversees all training programs. Its Basic Training Command is responsible for weeding-out candidates who should not be in naval special operations. That is why BUD/S, the 21-week-long initial training course for SEAL candidates, has such a high drop-out rate. "Hell Week" is not a mistake.

Advanced warfare training is offered by NSWC's Advanced Training Command. Throughout their careers, SEALs and SWCCs learn new skills and tactics that can give them an edge in combat, whether it's speaking Tagalog, or breaching a ship's steel door with an exothermic torch.

On a day-to-day basis, Naval Special Warfare personnel are involved in training and administrative activities with their assigned units. Over the past decade, WARCOM has streamlined its organizational structure to enhance operations. Today, there are six Naval Special Warfare Groups (NSWG), each with its assigned units as well as a specific region of the world identified for its area of operations. The latter allows the Group to become intimate with the terrain and weather, to acquire appropriate combat gear, and to develop combat strategies and tactics for that region in case they are ever needed.

Each Group is commanded by Navy Captain and has its own headquarters staff who are assigned to such responsibilities as intelligence, operations and planning.

★ **NSWG-1** (Coronado) manages SEAL Teams 1, 3, 5 and 7 on the west coast. This includes Logistics Support Unit 1, which provides the SEALs with combat gear and combat services (such as mobile communications). Since NSWG-1 responds to contingencies in the Pacific, Korea and the Middle East, it maintains a forward presence in Guam (NSW Unit 1) and Bahrain (NSW Unit 3).

★ **NSWG-2** (Little Creek, VA) oversees SEAL Teams 2, 4, 8 and 10 on the east coast, as well as Logistics Support Unit 2. This Group's area of responsibility is Northern Europe, the Mediterranean, Africa, and Central and South America. It has two forward-deployed units, both in Stuttgart, Germany: NSW Unit 10 focuses on the African continent, and NSW Unit 2 keeps an eye on the other regions. A detachment from NSW Unit 2 is also based in Panama City, Panama.

★ **NSWG-3** recently moved from Coronado, CA to Ford Island at Oahu, HI. It is responsible for all of Naval Special Warfare's undersea assets that are now consolidated at Pearl City, including SEAL Delivery Vehicle Team 1 (along with its mini-subs) and Logistics Support Unit 3. These assets deploy worldwide as needed. The Advanced Training Command's "mini-sub" school moved here too from Panama City, FL. The only undersea asset not in Hawaii is a detachment at Little Creek that helps maintain the Dry Deck Shelters used by SEALs to hangar their mini-subs on larger, host submarines.

★ **NSWG-4** (Little Creek) oversees the SWCCs and the high-performance boats they use to transport SEALs on missions. There are three Special Boat Teams: SBT-12 (Coronado) supports west coast SEAL Teams and is deployed to the Pacific and Middle East; SBT-20 (Little Creek) supports east coast SEAL Teams and is deployed to Europe, the Mediterranean and the Middle East; and SBT-22 (Stennis Space Center, MS) specializes in riverine warfare worldwide. NSWG-4 also manages a riverine-training program (NAVSCIATTS) at Stennis for allied, foreign military forces.

★ **NSWG-10** (Little Creek), serves as the intelligence, surveillance and reconnaissance arm of Naval Special Warfare. All intel systems, including remote senors, aerial drones and unmanned underwater vehicles, are managed by one of two Support Activities, which are staffed by cryptologists, intelligence analysts and communications experts. Support Activity 1 (Coronado) supports west coast SEAL Teams and Support Activity 2 (Little Creek) supports east coast SEAL Teams. NSWG-10 also staffs the Mission Support Center (Coronado), and manages the Cultural Engagement Unit. The latter is comprised of men and women who are linguists and regional/cultural experts. They accompany SEALs into the field on missions and help establish ties with the local inhabitants, as well as assist with intelligence-gathering activities.

★ **NSWG-11** trains, equips and oversees reserve SEAL Teams 17 (Coronado) and 18 (Little Creek).

And what about DEVGRU, the Navy's counterterrorist unit formerly known as "SEAL Team Six?" It's headquartered at Dam Neck, VA and is administratively attached to the Naval Special Warfare Command from which it receives supplies, support and access to training facilities. Both DEVGRU and the Army's "Delta Force" report directly to the Joint Special Operations Command (a component of SOCOM) and receive their orders from the President and the Secretary of Defense. On a day-to-day basis DEVGRU develops maritime, ground and aerial tactics that will benefit not only the SEALs, but America's special operations forces as a whole.

In addition to administrative and operational duties, each Naval Special Warfare Group is involved in research and development pertinent to its area of specialty. For instance, NSWG-4 focuses on emerging technologies that improve the performance and lethality of its combatant-craft. It is also interested in enhancing methods of inserting boats into combat, and quickly relocating them if necessary.

Among the many cutting-edge projects being pursued today by NSW Groups include: micro aerial drones for tunnel reconnaissance, structure-penetrating sensors, multi-fuel engines for boats, automatic 3D mapping of buildings, Heads-Up Display for SWCCs, stealthier boats with stronger lightweight composite armor, thermal diving gear that will allow SEALs to remain warm underwater for 12 hours, and a maritime mission-tracking system that can accurately monitor in real-time the location of surface craft, swimmers and divers within 5,000 meters.

This book is the first in the "United We Stand" series. It celebrates the SEALs and the entire Naval Special Warfare community for their accomplishments during the Global War on Terrorism by visually showing them in action at Sea, in the Air, and on Land. Please note that not every photograph is of a SEAL. Images of mission-support specialists who make the SEALs' combat successes possible are included as well. Spotlighting their contributions is the right thing to do.

SEALs: Naval Special Warfare in Action is intended for all military personnel and their families, military enthusiasts, government officials, civilians, and even our nation's youth. The last three audiences are important because society as a whole needs to better appreciate the sacrifices made every day by members of Naval Special Warfare and, thus, support their efforts.

To help with that learning process, care has been taken to ensure that the book's text and photo captions are easy-to-read. Military nomenclature and technical data are minimized whenever possible; the focus is shifted instead on the mission and the people. After all, weapons and gadgets don't make the warrior. Courage, determination, and spirit do. And the men and women of Naval Special Warfare have all three. *Hoo-yah!* ★

Right – Although SEALs prefer to maintain a low profile when deployed on a mission, sometimes it is necessary to use smoke to temporarily "blind" and confuse enemy forces. This gives the SEALs time to quickly scoot away and regroup (or attack if that is an option). To maintain communication among themselves, the SEALs rely on radio headsets like the one shown in this photo. The units are voice activated (only a whisper is needed), leaving the SEALs' hands free to carry and fire weapons. For ultra-quiet communication, such as when conducting a takedown of a terrorist inside a building, "throat-mics" are used. Affixed over a SEAL's larynx, the device allows him to whisper complete sentences without actually opening his mouth. The sensitive microphone picks-up the vibrations of the vocal cords and transmits the words.

Below – Using two Rigid-hull, Inflatable Boats (RIB) in a tandem formation, SWCCs practice a high-speed insertion maneuver. SEALs move from the RIB into the Zodiac (yellow craft), and then roll off the gunwale into the ocean, landing on their back. In combat, this technique is done with the Zodiac positioned on the open-sea side of the RIB. Doing so prevents anyone on shore from actually seeing the SEALs deploy and in what number. For extraction, the process is reversed, only the SEALs hook their arm into a large hoop and are pulled aboard the Zodiac.

SURVIVING
BUD/S

The year-long training required to become a SEAL is arguably the most grueling in the world. More than 70% fail. It's not unusual for some men to try two or three times before they finally earn the coveted Trident insignia. The selection, training and qualification process is intense because it has to produce highly skilled warriors who can be counted on to accomplish America's most sensitive "no fail" missions – often with strategic consequences – under significant stress.

A candidate must first pass the Physical Standards Test and C-SORT, a psychological exam that measures mental toughness. The two combined determine with 97% accuracy whether he will drop out by the end of Hell Week. Mental fortitude, the *"I will not quit, no matter what!"* attitude, is the determining factor that enables men to endure pain, fatigue and harsh conditions – in training and in combat. SEALs want a teammate who will not give up, who is trustworthy, who places team over self, and who will have their backs when things get tough.

The testing is followed by eight weeks at the Naval Special Warfare Preparatory School (Great Lakes, IL) and three weeks of Orientation. If the candidate completes both, he then attends BUD/S: Basic Underwater Demolition/SEAL. The Naval Special Warfare Center offers six classes a year at Coronado, CA that begin between February and October. The average class has 145 enlisted recruits and 15 officers, all of whom train together.

BUD/S lasts 21-weeks and is divided into three phases. Phase I focuses on physical conditioning and is marked by forced marches, long-distance runs, obstacle courses, boat drills, endurance swims, and log PT. Hell Week occurs during this phase and the men are subjected to five days of continuous training. Before starting Hell Week, the men swallow a small pill that monitors their core body temperature. The data is sent to medics using a handheld scanner. This monitoring is done to prevent and diagnose cold and heat injuries.

Phase II deals with basic combat swimming. They learn

The Making of a Warrior

Overleaf – The "drownproofing" evolution requires candidates to repeatedly sink to the bottom of the pool and then bounce back to the surface to take a breath of air – with their hands and feet bound. The drill, done for a half hour, teaches confidence and helps overcome fear of drowning. Photo by SF Tomajczyk

Left – Hoo-yah! SEAL candidates participate in a surf passage drill during Phase I of BUD/S. As a team, they will spend hours racing in and out of the ocean with the 108-pound "Itty Bitty Boat" (IBS - Inflatable Boat, Small) balanced atop their heads. By the end of BUD/S, many men will have a bald spot from the sand and friction.

about dive physics, hydrographic surveys, long underwater swims, and underwater knot-tying. Candidates undergo pressure-chamber testing to make certain they can safely dive to 300 feet, and are trained to use SCUBA and closed-circuit diving systems. They are subjected to "surf hits" whereby instructors suddenly attack and rip-off face masks and fins, and tie the regulator hose in knots. It's to teach them how to remain calm in an underwater crisis.

Phase III specializes in land warfare. The men are issued camouflage clothing and web gear, and undergo training in basic weapons, rappelling, navigation, small unit tactics, reconnaissance, and marksmanship. The second half of this phase is spent on San Clemente Island where they learn about explosives and demolition.

Upon completion of BUD/S, the men still have 26 weeks of SEAL Qualification Training ahead of them. During this time they learn close-quarters combat, ambush tactics, advanced weapons skills, small boat operations, and extensive land navigation skills. They also earn their parachute badge and spend a month at Kodiak, AK learning about cold weather operations (followed by 10 days of maritime operations in San Diego).

All candidates attend a 12-day Survival, Evasion, Resistance and Escape (SERE) class held in the mountains at Warner Springs, CA. It's a Level C course, meaning it is designed for military personnel who know secrets and who are at high risk of being captured. As such, the men not only learn survival skills but experience what it is like to be a prisoner of war, including being locked inside cages, and undergoing mock interrogations and torture.

At the end of SQT, the men receive their SEAL Trident, Special Warfare Operator (SO) rating and a Ka-Bar™ knife engraved with the name of a member of Naval Special Warfare killed in action, and they are officially assigned to one of eight active or two reserve SEAL Teams. (Note: SEALs who will pilot NSW mini-subs are assigned to SEAL Delivery Vehicle Team 1.) But their training isn't over. As they will discover firsthand, every day in Naval Special Warfare is a training day, and the only easy day was yesterday.

The SWCC training pipeline is similar to that of the SEALs, only it is 35 weeks instead of 58 weeks. Candidates partake in the NSW Preparatory School and Orientation sessions, followed by seven weeks of Basic Crewman Training at the SWCC Training Center of Excellence (Coronado) where they undergo physical training and learn basic seamanship, navigation, radio communications, small unit tactics, and swimming skills. During this time, they endure "The Tour," a 51-hour evolution similar to the SEALs' Hell Week, except it focuses on combatant-craft tactics and underway boat/swim events.

Candidates then advance to Crewman Qualification Training, which is broken into Basic and Advanced phases, each 7-weeks-long. The men become proficient in marksmanship (including vessel-mounted .50 caliber machine guns), insertion/extraction methods, casualty medical care, boat/propulsion systems engineering, and coastal/riverine patrolling techniques. Like SEALs, all SWCCs undergo Level C SERE training, and they are parachute qualified.

At the end of CQT, the men receive their "Boat Badge" and Special Warfare Boat Operator (SB) rating, and are assigned to one of three Special Boat Teams overseen by NSW Group-4.

An interesting graduation tradition is the "Compass Ceremony" held at the Vietnam Unit Memorial on Coronado. Each man receives a small wooden box from his future commanding officer that contains a compass. On the lid is etched *Mind Your Heading* and the name of a Vietnam-era sailor who served in the early boat units and paid the ultimate price. The SWCC Creed is inscribed inside the box. The men are told the compass binds them to the SWCC heritage and serves as a moral compass in their careers. They are also told to make their combat actions decisive yet measured, to always complete the mission, to never quit, and to leave no one behind. ★

Above – Pull-ups and push-ups are a normal part of the morning exercise routine during BUD/S. That's because upper-body strength is vital for a SEAL. After all, when deployed on missions, SEALs climb caving ladders, rappel buildings, climb walls, scramble over obstacles, fast-rope from helicopters, etc. while carrying upwards of 100 lbs. of tactical gear.

Left – Covered with irritating sand and resembling a sugar cookie, SEAL candidates perform a myriad of physical-strength exercises using a log that weighs between 400 and 600 lbs. The men, all in the same boat crew, will do group sit-ups with the log resting on their chests, or hold it over their heads while marching down the beach on a long hike. If they do not work as a team, the log will slip from their hands, causing possible injury and earning them additional exercise from the instructors as punishment. The worst is to be given "Ol' Misery," the largest and heaviest log of them all. The words "Misery Loves Company" is carved into its side. Allegedly, one BUD/S class stole and tried to torch it, but the log refused to burn.

Left – The obstacle course at BUD/S offers many challenging stations, including this unstable rope net, that test strength, agility, balance and speed. SEAL trainees are expected to improve every week, and to exceed the minimum time posted by the instructors.

Right – If a candidate wants to quit BUD/S, all he has to do is ring this brass tugboat bell three times and place his helmet on the ground. The bell was a gift from Class 58 to the program in 1970. Although it is mounted adjacent to the "grinder" at the Naval Special Warfare Center, the instructors bring it along when the men do long marches or swims. Out of every 100 men who enroll in BUD/S, upwards of 80 end-up ringing the bell. The activity that causes the bell to be rung most is "temperature conditioning" evolutions whereby the men spend considerable time either wet or in the ocean. Water draws away body heat 25% faster than air, leaving candidates shivering uncontrollably as their body's core temperature drops.

Below – Surf passage drills may look like fun, but strong currents (including rip tides) and pounding waves can create a nightmarish situation. Paddles are lost, boats overturn and men can be swept out to sea. The drills, which are conducted both during the day and at night, are a prelude to formal training later on using the SEAL Teams' Zodiac combat boat.

Candidates aspiring to become SWCCs, crawl through the surf during the final training evolution known as "The Tour." It is a three-day session that pushes the men to their physical and mental breaking point. Operating in the same environment as SEALs and directly supporting their mission, crewmen must learn to endure prolonged exposure to rough seas and cold, wet conditions. Each SWCC class has about 50 candidates, half of whom fail to pass.

Above – At the end of Hell Week on "So Sorry Day," the candidates are exposed to simulated combat conditions. Crawling beneath barbed wire in the sand pit shown here, the men participate in "whistle drills." Every time the instructor blows the whistle, they quickly cover their head and ears, open their mouths, and cross their legs. Although it may seem strange, in reality there is a purpose for this drill. In combat, this position helps protect them from the blast wave of a nearby exploding grenade, bomb or improvised explosive device. The instructors use smoke and artillery simulators to make the training more realistic.

Right – A candidate crawls through smoke, barbed wire and obstacles during a training scenario. The concrete tube, which is barely shoulder wide, is one of many that lead into and from a slimy, water-filled mud pit. At this point in Hell Week, the men are burning 7,000 calories a day and have only had four hours sleep in five days; many are hallucinating. Those who endure Hell Week earn the right to wear a brown T-shirt, which signifies their accomplishment.

Right – SWCC candidates perform a "dump boat" exercise with a Combat Rubber Raiding Craft during Basic Crewman Training. In their support role to the SEALs they must know how to recover from at-sea incidents, such as righting a flipped boat and getting it operational as quickly as possible. In combat, they easily could be under enemy fire where every passing second means the difference between life and death.

Below – In Phase II of BUD/S candidates focus on learning diving skills, the very essence of being a SEAL. This photo shows a diver participating in a "night gear exchange" drill. With his diving mask blackened out, he must use his sense of feel to trade his diving equipment (e.g., dual air tanks, regulator) with another diver, put it on and have it working within just a few minutes. The drill provides valuable skills that will be used in the future on missions where SEALs conduct an underwater approach on a target, such as a ship or oil platform. Likewise if they have to suddenly trade-out a malfunctioning air tank for some reason.

Rappelling is used to quickly descend cliffs, climb down the outside of a building during an assault, and infiltrate from a helicopter hovering 60-100 feet above the ground. SEALs learn rappelling basics during SQT in which they practice techniques from a 60-foot-tall tower (shown here). The most common mishap a beginner makes is known as the "opossum," when he flips upside-down because he doesn't properly use his brake hand. Naval Special Warfare uses the military standard 11mm, 150-foot-long kernmantle static-rope for rappelling operations. It has a breaking strength of 7,650 lbs.

Left – A SEAL candidate wades ashore San Clemente Island ("The Rock") during an over-the-beach exercise. The training prepares him to conduct the SEALs' signature mission: Operations that begin in the ocean and transition to land. Similar training is conducted at the Silver Strand Training Complex where SEALs practice hydrographic and photo reconnaissance skills, and clandestine movement from the surf, inland.

Right – In combat-swimmer training the men learn the art of underwater navigation using an "attack board," which features a compass, watch and depth gauge. Using a consistent fin stroke and counting kicks, the diver knows how far he has traveled. (An eight-man element covers 100 yards in an average time of four minutes.) The diver with the board ("Compass Man") is responsible for accurately leading the group to their target, on time. The 10 x 12 inch navigation board shown here measures depth to 200-feet while simultaneously tracking mission and leg time. The compass dial "glows" for up to 8 hours.

Below – A candidate in the diving phase of BUD/S, participates in a simulated casualty drill. The red smoke requests immediate medical evacuation by helicopter. All SEALs swim in dive pairs for safety, and are linked to one another and to the larger swim element with a buddy line.

SWCC trainees involved in the advanced Crewman Qualification Training secure an injured victim to a spine board during a medical training scenario. On real-world missions, SWCCs provide medical care to SEAL casualties in addition to their normal duties of transporting SEALs in the Navy's high-speed boats. The Navy's SEAL and SWCC Scout Team keeps an eye out for athletes who participate in sports like water polo, triathlon, rugby, lacrosse and swimming since they tend to be suited for Naval Special Warfare. These sports involve both a team and individual effort, and they require persistence and determination.

Above – During the last phase of BUD/S candidates learn about land navigation, demolitions, and small unit tactics. After BUD/S they move on to SEAL Qualification Training, which provides tactical skills they will need to join a platoon. And after passing SQT and officially becoming a SEAL, they begin an on-going cycle of advanced individual and unit-level training for the rest of their careers. The first course is three months of language and cultural training.

Left – During maritime operations training candidates learn how to position themselves aboard a 15-foot-long Combat Rubber Raiding Craft so they keep a low silhouette, as well as ensure they are not tossed overboard. The boat is loaded so its center of gravity is one-third from the stern to maximize its stability, speed and performance. The Zodiac takes just two minutes to inflate using an air tank, and it is crewed by a coxswain (seated on the port stern) and an engineer (who maintains contact with other Zodiacs in the insertion or extraction formation).

Left – Land navigation is a critical skill SEALs need to master since they are routinely inserted miles from a target and have to approach it by way of forests, deserts or mountains. After learning how to use a map and compass, candidates spend time in the rugged mountains surrounding Camp Michael Monsoor (NSW Mountain Warfare Training Center) and nearby Camp Morena putting their knowledge to use navigating different routes over various terrain, including lake and riverine obstacles. Some courses are 5,000-meters long, and each has a time requirement the men must meet or exceed.

Right – After completing BUD/S, SEAL candidates are enrolled in an intense four-week-long parachute course that qualifies them in static-line and free-fall jumps. The self-inflating "ram-air" parafoil shown here gives the men control over speed and direction. When used on high-altitude, high-opening (HAHO) jumps at night, SEALs can quietly glide more than 30 miles with low probability of being detected.

Below – During SQT, candidates perform buddy carries between stations during a 36-round shooting drill at Camp Pendleton. Their marksmanship skills using the M4A1 carbine are tested at 100-, 200- and 300-yards. Not an easy thing to do when you've been winded carrying a 200 lb. "casualty."

Candidates in the final phase of BUD/S use a rope to guide themselves down a cliff and into the ocean during a nighttime field exercise on "The Rock." Each candidate puts all the skills he's learned over the months to use, trying to convince the instructors he has what it takes to be a Team Guy.

Above – As part of close-quarters combat (CQC) training, SEAL trainees burst into a room and engage hostile targets found in their assigned sector with precise, short-bursts of gunfire. The men receive three weeks of CQC skills during SQT at the Mountain Warfare Training Center (Camp Michael Monsoor), as well as at the Silver Strand Training Complex. They undergo more advanced training once they are assigned to a SEAL Team. CQC techniques are often used by SEALs since their missions usually involve tight quarters (ships, buildings) and they engage targets at close range (within 30 meters).

Left – With the Pacific Ocean glowing under the setting sun, SEAL candidates head for an offshore oil platform where several hours later they will board and assault it under the cover of darkness. Typically, the Zodiac drops the men 500-1,000 meters downwind from the oil rig so that its engine cannot be heard. The men swim the rest of the way and use a pole to put the caving ladder in place. Maritime operations like this are taught to the trainees during SQT.

Left – A SEAL trainee practices room-clearing tactics in the "kill house" located at Camp Monsoor. The multiple rooms and hallways in the facility have movable walls, allowing for different configurations. The blue dots seen on the ballistic wall are made by Simunition, a cartridge that features a plastic bullet filled with colored powder. It leaves a bright mark on impact, enabling instructors to critique a shooter's accuracy and technique. CQC training is progressive: it begins with the trainees shooting blanks (or lasers), then Simunition and, finally, live ammunition.

Right – Knowing how to destroy something with explosives is a skill that dates back to World War II when Navy "frogmen" planted limpet mines on enemy ships and blew-up defensive obstacles on beaches prior to amphibious landings. SEAL trainees learn about - and practice using - military explosives (C4, PBXN, bangalore torpedo, hose charge), initiators, mines and breaching charges at San Clemente Island, as well as at the Silver Strand Training Complex.

Below – A flare illuminates a SEAL trainee during a nighttime training exercise on "The Rock." The men, as a squad, learn how to conduct reconnaissance and combat patrols in the dark, as well as how to quickly disengage the enemy should they unexpectedly walk into an ambush or come under enemy fire.

FROM THE
SEA

Attacking from the sea in small numbers is the SEALs' specialty. They own the ocean and it is what separates them from all other American special operations forces. Missions, which can extend well over 50 miles, involve swimming and the use of surface craft, helicopter, submarine and/or mini-submarine.

The SEALs have their own "secret Navy" to carry out maritime operations. At one end of the spectrum – the small, high-speed craft – there is the current RIB and SOC-R, as well as the newly introduced High-Speed Assault Craft and SEALION. (All are operated by SWCCs in the Special Boat Teams.) And at the other end of the spectrum – large warships – there is the *Cyclone*-class coastal patrol ship, which is augmented by the High Speed Vessel and the new *Freedom*- and *Independence*-class littoral combat ships. NSW also uses leased commercial ships like the MV *C-Commando* to discretely transport SEALs and their mini-subs around the world.

For enhanced underwater-swimming approaches, SEALs use the Swimmer Transport Device (a type of underwater scooter) and JetBoots. The latter is a miniature water-jet propulsion system that straps-on to a diver's thighs, propelling him along at 4 knots to depths of 300 feet. His hands are left free to carry weapons.

For long-range, underwater missions requiring stealth, the SEALs turn to *Los Angeles*- and *Virginia*-class attack submarines that can drop them off near the coastline. They Lock In/Lock Out underwater via the escape trunk.

The Navy has six Dry Deck Shelters to transport SEAL mini-subs. Four are attached to the decks of *Ohio*-class guided-missile submarines (SSGN) that have been converted for special operations missions. The other two can be installed on attack submarines if needed.

These units, which cost $30 million each, are essentially a 9-foot-diameter soda-can shaped "garage" attached to two 7-foot-diameter spheres. The garage is where the mini-sub (or unmanned underwater reconnaissance vehicle or multiple Zodiacs) are stored. A clamshell door opens and closes to the ocean. SEALs gain access to the hangar via a sphere

Maritime Assault

Overleaf – *You never know when you'll bump into a Navy SEAL. This one is armed with an MP5 submachine gun. The Mk-25 rebreather the combat swimmer is using guarantees there are no bubbles to alert anyone of his presence. Only the fish know, and they're not talking. – Photo by SF Tomajczyk*

Left – *Members of SBT-20 transport SEALs in an 11-meter-long, rigid-hull inflatable boat (RIB). The armed craft, which hits speeds of 40+ knots, is crewed by three SWCCs and is designed to insert/extract a squad of SEALs in medium-threat environments. It's equipped with a gyro-stabilized FLIR system (to "see" at night) and a laser range-finder and pointer.*

latched onto the submarine's escape trunk. The other sphere is a hyperbaric chamber used in medical emergencies. The spheres and hangar are enclosed within a fiberglass fairing to form a single unit that slices neatly through the water.

Since 1995, the SEALs have used the Mk-8 MOD 1 SEAL Delivery Vehicle as their mini-sub. It was designed and built by Coastal Systems Station, a Navy R&D facility. (Today it's known as Naval Support Activity, Panama City.) Referred to as the *Gator*-class, the 20-foot-long, 5-foot-diameter subs are piloted by two SEALs and can transport another four passengers (six if they have no gear) in excess of 35 miles. Weighing 6,000 lbs., the SDV slides in and out of the DDS hangar on an extending track-and-cradle system (which can be jettisoned in an emergency so the host submarine is not placed in danger).

The SDV is a "wet" mini-sub, meaning it is flooded by seawater. Hence, SEALs breathe from on-board oxygen flasks and wear diving suits and active thermal gear (6.6v dry suit liner and gloves) to stay warm. Although everyone is enclosed inside the Mk-8, piloting is facilitated by the use of a Doppler Inertial Navigation System, forward-looking obstacle avoidance sonar, and a mission support computer. The men communicate with each other using the intercom system. To peek above the surface, the SEALs extend a collapsible mast and look through a 360-degree periscope (which also has a camera).

Ten SDVs remain in service today, two of which were modified in 2010 to travel at depths deeper than originally designed. In spite of the Mk-8's success, it is at its life end and is now being replaced on a one-for-one basis by the Mk-11 Shallow Water Combat Submersible (SWCS). The retired SDVs are being looked at as possible test platforms for the "Sea Predator" program, an unmanned drone armed with weapons and sensors.

SWCS, built by Teledyne Brown Engineering, is 22-feet-long, weighs 10,000 lbs., and is six-inches larger in diameter than the SDV. The increased size meant the Navy had to reconfigure the interior of the DDS to accommodate the new mini-sub, which even now barely squeezes inside.

SWCS uses the Voyage Management System for mission planning and analysis, and offers command and control displays for the pilot, navigator and one crew member. It features an advanced reconnaissance system comprised of electro-optic and infrared sensors, and has enough air on board to meet a minimum 8 hour, 45 minute mission transit time requirement.

Like the Mk-8 SDV, the Mk-11 is wet. However, the Navy is introducing a Dry Combat Submersible (DCS) for the SEALs. There are two versions: "S351" (UOES-2) and "Button" (UOES-3). The S351 is built by Submergence Group. It is 39-feet-long with a roughly 7-foot diameter, and weighs 57,300 lbs. It is piloted by two SEALs and can transport six passengers 130 miles at depths of 375 feet.

"Button" is built by General Dynamics. It is 31-feet-long with a 6-foot diameter, and weighs nearly 37,500 lbs. It is piloted by two SEALs and can transport four passengers 60 miles at depths of 190 feet. The Navy intends to redesign and lengthen the DDS to hangar this particular mini-sub.

To enhance its undersea warfare capabilities, NSW in recent years consolidated all its undersea assets in Hawaii. Pearl City is now home to SDVT-1 and the "schoolhouse," as well as the mini-subs. NSW Group 3, which oversees all these, is headquartered in Bldg. 55 at nearby Ford Island. A detachment that maintains the DDSs is located at Little Creek, VA. This unit also supports mini-sub training on the east coast and at Key West, FL.

For cold-water operational training, SDVT-1 travels from Hawaii to the Keyport and Dabob Bay ranges in Washington. The six-weeks of training culminates with a mock mission in which a mini-sub stealthily travels from Port Townsend and delivers 4-6 SEALs at Indian Head. The SEALs conduct special reconnaissance on the island and are extracted by mini-sub two days later. ★

This is a scene you won't see any longer. The Mk-V (SOC) high-speed boat (top) – which hit speeds in excess of 50 knots – was retired by Naval Special Warfare in late 2012, and the RIB (bottom) is now on its way out as well. Both are being replaced by a new family of armed combatant craft.

Above – Members of a SEAL Delivery Vehicle Team practice boarding a submarine by fast-roping from a MH-60S helicopter. Submarines provide SEALs with an ideal way to travel long distances and get close to an enemy target without being detected. Although this attack submarine, the USS Toledo, does not have a Dry Deck Shelter to transport the SEALs' mini-sub, the SEALs can still conduct recon and other missions by either swimming or using a Zodiac. They are trained to egress underwater through the escape trunk (shown here). If you're curious about the pattern on the hull, those are anechoic tiles. Made from synthetic polymers, they absorb sonar from enemy ships and aircraft, making the submarine nearly undetectable.

Right – Zodiac F470s zip-up onto a beach to insert combat-armed SEALs. The boats have eight airtight chambers and are powered by a 55 hp engine with a shrouded impeller, which reduces injury risk to SEALs when they are in the water. It also limits damage to the propeller if a rock or underwater obstacle is hit.

Top – Members of Special Boat Team 20 practice their marksmanship with the Mk-19 grenade launcher by firing at a tanker sunk near Piney Island, NC. The weapon fires 40mm grenades up to 2,000 meters.

Above – The Cyclone-class coastal patrol ships are dedicated Naval Special Warfare "mother ships" from which a squad of eight SEALs can operate using Zodiacs and even the Mk-8 mini-sub. They are being replaced by the larger Littoral Combat Ship.

Right – A SWCC coxswain at the helm of his RIB makes final preparations before heading seaward.

A SEAL assault team practices an underwater stack drill at the Combat Training Tank. Note that the swimmers are connected to each other by a "Lizard Line." This ensures they do not get separated or left behind, a distinct possibility at night or when swimming in murky water. The pole a SEAL is carrying (center) extends and is used to put a wire ladder in place on the ship or oil platform so they can board it.

Above – To move quickly underwater – without a long, tiring swim – SEALs use this lightweight, collapsible Swimmer Transport Device. The unit can transport two combat swimmers and 100 pounds of additional gear at speeds of 3+ knots for more than two hours, and to depths in excess of 80 meters. The STD features moving-map navigation and bottom scanning sonar. Digital displays provide data on position, heading, course, speed, depth, and routing waypoints and tracking. For SEALs, the unit means less time spent in cold water during an insertion from a submarine. It also means they can cover more territory during reconnaissance missions.

Right – A four-man swimmer scout team moves ashore 100-300 meters ahead of the main SEAL assault element. They will quickly cross the beach and move inland as a group, separate, and deploy left and right to conduct a close-in reconnoiter – searching for dangers and enemy presence near the beach landing site. Note they have removed their fins but are still using their UBA in case they need to abruptly retreat.

Right – SEALs practice a ship-boarding technique using a hook-and-ladder in the Combat Training Tank. Note that the grapple hook is wrapped in black tape to prevent glint, as well as to dampen any noise it might make when put into place. The hook features a "D" loop designed to keep the aluminum ladder off the ship's hull so it doesn't scrape and alert the crew to the SEALs' presence. Climbing a caving ladder when it is free-hanging is difficult since the 6-inch wide rungs are attached to flexible wires. SEALs hold on to the rungs using over- and under-hand grips and they wrap their legs <u>around</u> the ladder so their heels go into the rung from the other side. The ladders come in 5 and 10 meter lengths that can be linked to each other.

Below – Camouflage is routinely used by SEALs. Face paint is used to cover all exposed areas of skin, including the hands, face, ears and neck. Parts of the face that naturally create highlights (nose) are darkened, and those that form shadows (eye socket) are lightened. By doing this, human vision is fooled since we are accustomed to seeing it the other way. Hence, recognition is slow or even nonexistent. Camouflage is applied in stripes, blotches or using a combination of both. The colors used depend on where the SEALs will be operating. For instance, the SEAL shown here is wearing clothing and face paint that is appropriate for beach and desert environments.

Left – Swimmer scouts serve as the advance team for the main body of SEALs. They swim ashore from outside the surf zone and recon the area for threats, as well as locate a beach landing site, an assembly area, and cache sites to hide gear. A small microphone in the closed circuit rebreather's mouthpiece allows for diver-to-diver communication.

Right – A combat swimmer is typically equipped with a wet suit, face mask, UBA with mic, fins, life vest, Special Operations M4A1 carbine, sidearm (SIG Sauer P226 , P228 or Heckler & Koch 45C with optional suppressor), and a knife. The SEAL shown here is also equipped with a basic navigation board. The newest model, the Navigator, features sonar imaging (250 meter range) and a Doppler navigation system.

Below – SEALs use the 30-lb. chest-mounted Mk-25 rebreather, which runs on pure oxygen. Every breath is filtered by soda lime to remove carbon dioxide. No tell-tale bubbles are released in the process, making the Mk-25 ideal for clandestine missions. Divers can swim for four hours at depths up to 25 feet. Swim time sharply decreases with depth (10 minutes at 50 feet) since oxygen begins having a toxic effect on the body. For deep dives the SEALs turn to the 70-lb. Mk-16 rebreather, which uses mixed gases: Nitrox (to 190 feet) and Heliox (to 300 feet).

Above – A SEAL helo-casts from a MH-60S Knighthawk helicopter assigned to the "Chargers" of Helicopter Sea Combat Squadron 26. Helo-casting is a quick insertion method in which SEALs jump from a helicopter as it travels 10-feet above the waves at 10 knots. The forward motion ensures the men do not land on top of each other. Zodiac boats, if needed, are fully inflated ("Hard Duck") and attached to the underside of the Knighthawk and dropped first. On missions, helo-casting is often done at twilight 20-miles offshore. This allows the SEALs to see what they're doing. By the time they reach the beach, they are under the cover of darkness.

Right – A member of SEAL Team 5 scans the rugged shoreline for threats as he and the rest of his fireteam practice a beach incursion during Exercise Northern Edge. Northern Edge is an annual, joint-service field exercise that takes place in Alaska and involves more than 5,000 participants. During the two weeks the military services learn how to integrate better.

Right – After swimming ashore, a SEAL fireteam swiftly, quietly and with intent moves inland using the rocky terrain for concealment, as well as protection from gunfire. Note they remain in dive pairs and are simultaneously moving in a "leapfrog" manner in which one SEAL in each dive pair provides cover with his weapon while his partner advances to a new position. (Each advance is a 3-5 second rush.) When he arrives, he then provides cover for his partner. This bounding movement ensures the SEALs are ready to immediately engage the enemy.

Below – SEALs are taught to use the element of surprise. Here they are hugging the shoreline of San Clemente Island, using the rocks and cliffs to conceal their movement. Few people would ever suspect their presence. On a mission where a SEAL platoon inserts by RIB, they do so at night during high or slack tide to reduce the risk of being discovered. They stop and form a "boat pool" 2,000 yards from shore. The lead RIB then moves closer and launches the swimmer scouts to recon the landing site. Once the "All clear" signal is given, the RIBs advance as far as surf conditions allow and insert the SEALs. The north end of San Clemente is reserved for Naval Special Warfare. It includes weapon ranges, a 26-building mock city, and an airfield. Here, away from prying eyes, SEALs conduct tactical exercises in realistic conditions.

A SEAL assault team ("The Train") gets into position to clear a ship compartment. The men stack outside the hatch with the Number One man (left) providing door security and the last man providing rear security. When ready to begin the assault, each SEAL gives a squeeze signal to the man in front of him. In this photo, the second SEAL is giving that silent message.

Left – After coming ashore, members of SEAL Team One gather intelligence using a range finder (right) and an underwater digital camera. The latter, a Nikonos that has been specially modified by Kodak for the SEALs, links with a field transmitter to send images to military decision makers in near-real time.

Below – A combat swimmer practices placing a small inert magnetic mine on a ship. On real missions the Mk-5 limpet mine is often used. The size of a garbage can, it holds 100-lbs. of explosive that can break the spine of a ship. It's held in place against the hull by two tube-shaped air bags, one on each side. SOCOM seeks to develop an improved limpet mine that is 50% smaller and lighter, yet 2-3 times more powerful.

Right – SEALs practice boarding an anchored ship. They surfaced directly beneath the ship, hugging the contours of the hull to conceal their presence. A three-man security team (right and bottom) is keeping an eye out for anyone who might approach. The other men have removed their swim gear and attached it to a derigging line and are now preparing to climb the ladder. The Point Man (left), armed with a handgun and Special Operations M4A1 carbine, recons for any threats. If all is clear, he will board and maintain 360-degree security while the Officer-in-Charge and other SEALs climb up.

Left – A SEAL confidently climbs the narrow ladder that links the bouncing RIB to the oil rig. Falling is not an option. Take note of the SEAL's "Secret Squirrel" shoulder patch. Named after the children's cartoon show from the 1960s featuring a squirrel as a secret agent, it's often worn by those in the military who are involved in clandestine missions.

Right – SEALs and members of Special Boat Team 12 practice covertly boarding a gas and oil platform from a RIB. For SEALs wearing full assault gear, the climb can be strenuous and difficult, especially in rough seas and at night.

Below – The Navy uses bottlenose dolphins and sea lions to patrol harbors, recover objects, locate mines, and search for suspicious objects. Zak, a 375-lb. California sea lion, is part of the Mk-6 System. (It is one of five marine mammal systems. Each is designed for specific missions.) Zak is trained to detect intruders and swimmer delivery vehicles, and then alert security forces (which includes SEALs). The leash the SEAL is holding in his left hand in this photo runs through a link attached to the Zodiac's floorboard. If Zak becomes aggressive trying to reach and eat from the chum bucket, the SEAL yanks on the leash. In so doing, Zak's head is quickly pulled down and away, preventing the SEAL from being seriously bitten.

Left – Aerial snipers are invaluable to shipboarding operations. On such a mission two SEAL snipers are deployed, each in his own helicopter. They link-up with the airborne SEAL assault team five miles astern of the ship. Serving as escorts, all helicopters then speed toward the target at just 50-feet above the ocean. Upon arrival, the snipers peel-off left and right and provide cover to the assault team as they fast-rope onto the deck of the ship.

Right – Naval Special Warfare has a 3,500-acre training range along the Pearl River (Stennis Space Center, MS). SBT-22 is headquartered there in a 15,000 sf Riverine Operations Center, where it develops riverine combat tactics, such as extracting SEALs under enemy fire. Using shallow-draft boats, they approach from the ocean and speed up the river to the extraction point. Smoke conceals their actions. SBT-22 also trains at the Salt River Range at Fort Knox.

Below – The SOC-R is a heavily armed riverine craft since most hostile engagements occur at close range, in confined space. Weapons include miniguns, grenade launchers and machine guns. This photo also shows a radar screen (center), swimmer platform on the stern, and a multi-tube smoke-canister dispenser (left). The SOC-R features waterjet propulsion and a low-drag hull that allows the agile craft to top 40 knots.

The Navy converted four Ohio-class ballistic-missile submarines to support clandestine missions. Each sub berths 66 SEALs and features a virtual-reality weapons range. Two of its 24 missile silos have been turned into diving chambers, and the remainder are used to launch Tomahawk cruise missiles and store the SEALs' gear, aerial drones and specialty packages. The Dry Deck Shelter hangars the SEALs' mini-sub (or an unmanned intelligence-gathering submersible like the Sea Stalker, SHARC, SAHRV, or Littoral Battlespace Sensor). A high-tech Combat Management Center within the sub enables SEAL missions to be planned and, once begun, monitored in real-time.

Left – Submarines often secretly transport SEALs on missions, especially maritime operations. Getting aboard a sub that is on patrol at sea, though, requires a bit of help. Shown here, an Army MH-60 special operations Black Hawk flown by the elite "Night Stalkers" makes a delivery. The SEALs fast-rope to the stern and then dash for the open hatch of the Dry Deck Shelter. (The photographer of this image was inside it at the time.) The DDS hangars their "Eight Boat" and provides access to the submarine's interior.

Below – After exiting the submarine through its aft escape trunk (shown at center), SEALS prepare to inflate and launch a deflated Zodiac boat ("Rolled Duck"). On subs that do not have a Dry Deck Shelter, such as this one, the boats and submersible outboard engines are stored either in a sail locker or within the submarine itself in the torpedo room.

Right – Locking-out of a submarine can be tense and claustrophobic, especially on older subs that have small emergency-escape trunks. In those, five SEALs and their gear crowd into the 56-inch diameter trunk and then flood it. They remain locked inside until the pressure equalizes with the ocean. By contrast, the new, fast-attack Virginia-class submarines (shown here) have a large, dedicated diving chamber that allows up to nine SEALs to exit or enter at a time.

Above — Launch and recover operations for the Mk-8 take place with the submarine at a hover, moving at 1.5 knots at a depth of 30-90 feet (max. 130-feet). This photo shows the mini-sub's pilot and navigator compartment (with sliding door open). Displays provide data regarding the Doppler Inertial Navigation System, GPS, underwater mapping, air supply, forward-looking obstacle avoidance sonar, mission support computer, and the rendezvous and docking system. A collapsible mast mounted on the top centerline features a periscope and radio antenna.

Right — An 8-man crew is in position to recover a SDV. Some are breathing from "hooka" air lines from the shelter while others use Mk-16 rebreathers. Each man has an assigned role, which enables the Hangar Supervisor to know if someone is out of place or doing something wrong. Two divers are standing inside the "clamshell," the 9-foot-diameter door that weighs 8,000 lbs. It is opened and closed by a hydraulic system and is secured shut with 18 swing bolts.

Right – A view of the aluminum track-and-cradle system fully extended from the Dry Deck Shelter, ready to receive the Mk-8. The mini-sub uses a pinger locater to home-in on a transducer affixed to the buoy (top), connects to the buoy line, and is then winched down to the cradle. After being secured in place, the track is retracted by a winch, thereby transporting the Mk-8 inside the hangar. The track system, which is deployed manually by two divers, was recently strengthened to support an additional 10,000 lbs. In so doing, it can accommodate heavier warfare payloads (e.g., UUV, mines, sensors), as well as the new Mk-11 Shallow Water Combat Submersible (SWCS). The next generation DDS will likely feature a remote-control track-and-cradle system, removing divers from the process and limiting injuries.

Below – An older view of the SEALs' original Mk-8 during training that shows a dual DDS arrangement on a sub. Today, the Navy has six shelters deployed on submarines. Each measures 40-feet in length, weighs 68,000 lbs. and is made up of a hyperbaric chamber, an access trunk (leading into the submarine), and a 26-foot-long hangar. While the hangar easily handles the Mk-8 MOD 1, the new Mk-11 SWCS barely fits inside. The Navy had to rearrange valves and winches near the internal control station, as well as move the hydraulic system outside the hangar.

Left – Traditionally, the SDV pilot is an enlisted man; the navigator, an officer. They are seated side-by-side in the bow of the Mk-8 and plug themselves into the mini-sub's various systems, including the air supply. The boat has 88 hours of air on-board, which sounds like a lot until you divide it among the number of SEALs seated in the sub. The Mk-8 can transport up to eight men, meaning there is actually only 10 to 11 hours of usable air. Most underwater missions last 6 hours or longer. (The record is allegedly 14 hours.) That's a long time to be sitting in cold water. The SEALs are exploring technologies to keep warm, including the use of aerogel-lined diving suits and hydrogen catalytic heaters that circulate warm water through a network of tubes on the diving suit.

Below – Crewmen wear the Mk-20 full-face mask with a built-in microphone that allows them to communicate with each other and passengers while in transit. To communicate long-distance with the host submarine, surface ships and even shore commands (using a gateway), the Mk-8 is equipped with HAIL. It's a hydro-acoustic system that provides discrete, jam resistant, through-water communication, including the ability to send and receive email and text messages underwater. HAIL signals can also be used by the crew to determine the distance between the SDV and the platform they are in contact with.

Above – SEALs enjoy the sunset after conducting SDV training operations. The USS Michigan (shown here) was originally a ballistic-missile submarine and is now a dedicated special-operations sub. Three others join her. All of them underwent modifications to host the DDS, including mating hatch revisions; addition of electrical penetrations, valves and piping for ventilation; air for divers; draining water; additional ballast; and changes to the submarine's hovering control system.

Left – On real-world missions the typical SDV profile consists of a deep launch from the host submarine (90-130 feet), a shallow transit for much of the journey (30 feet) – even surfacing at times ("turtling") to conserve air – followed by a deep approach near the coast. The SEALs then "sink" and anchor the Mk-8 mini-sub to the ocean bottom and swim ashore to complete their mission. Once done, they use a hand-held pinger locater like the "Dive Ranger" to find the sub by homing-in on a transponder mounted to it.

Members of SEAL Delivery Vehicle Team 1 practice being extracted from the ocean by an Army Black Hawk helicopter using a SPIE rig. The men, each wearing a harness with an attached carabiner, hook themselves to a D-ring woven into the heavy-duty rope. (The ring pairs are spaced seven feet apart on the line.) The helicopter then lifts and transports them all together. Once free of the water, the SEAL partners link their arms and legs to create a more aerodynamic flow and prevent spinning while in flight.

FROM THE AIR

Because SEALs are a maritime force, people don't think of them when it comes to aerial assault. That's actually a good thing, for it allows SEALs to take advantage of that ignorance and successfully attack the enemy by parachute and helicopter.

SEALs learn how to parachute right after BUD/S, just as they begin SQT. In the past, they attended the Army's schools in NC and AZ, but now Naval Special Warfare has its own parachute school, which is run by Tactical Air Operations.

The static-line course is first. It teaches the basics of jumping from a plane using the MC-6 parachute and landing. It essentially condenses the Army's 21-day class into five days, with each man making five jumps into the "Monsoor Drop Zone" at Nichols Field in San Diego.

This course is followed by the free-fall class, which is three-weeks long and is broken into three phases. The first phase consists of passing eight free-fall proficiency categories. Each category teaches specific skills and builds on the one before it.

For instance, Category A focuses on topics like body position, canopy control and landing principles. Categories B thru H sequentially build on that by teaching such skills as tracking, adjusting fall rates, water landings, and braked turns, approaches and landings. The men are given three attempts to pass the requirements of each category.

The second phase transitions to military free-fall operations. It teaches SEALs how to make high-altitude parachute insertions using the MT2-XX and brand new RA-1 "ram-air" parafoils. Training includes the proper rigging of combat gear and weapons, military hand signals, group parachute formations, and navigation. With regard to the latter, SEALs have relied on altimeters, compasses and wrist-mounted GPS systems, like the Garmin™ 401. However SOCOM is developing a common parachute navigation system for all SOF units to use. One potential design is a helmet with built-in GPS. On the inside of the visor, a heads-up display is shown that provides a moving map and key data, such as direction, distance and time to the target, current altitude, and ground speed.

Aerial Assault

Overleaf – *A SEAL parachutes into the frozen tundra of Norway. SEALs assigned to NSWG-1 and NSWG-2 train for combat in the Arctic and cold regions of Europe, Asia and the north Pacific. They learn that frozen lakes and rivers may be the best drop zones, and that landing backwards in deep snow may save them from a broken leg.*

Left – *A Seahawk helicopter from Helicopter Combat Support Squadron 6 (HC-6) provides suppressing fire in support of SEALs conducting an urban-assault training operation in a mock city at Fort Knox.*

During this second phase the men also learn about atmospheric physics and medical hazards (hypoxia, hypothermia), and how to use supplemental oxygen systems. Prebreathing pure oxygen is required before flying above 10,000 ft., or risk getting the bends. The amount of time is determined by the altitude and type of jump to be made. For instance, a high-altitude, low-opening jump from 25,000 ft. requires 30 minutes of prebreathing, whereas a high-altitude, high-opening jump requires 45 minutes.

After mastering the military phase, the men finally perform night jumps. It begins with a "slick jump" (no equipment), followed by an oxygen jump, and then by two "combat jumps" (oxygen, rucksack and weapons). When jumping, the men follow the leader out of the plane and gather around him in a Wedge or Column formation. A small chemstick is affixed to the back of everyone's helmets so the men can find each other. If night vision goggles are worn, infrared chemsticks are used instead.

Students graduate the free-fall course with a total of 30 jumps. That's just enough knowledge and experience to be dangerous, which is why NSW sends them on to advanced parachute training courses, and requires that every SEAL make at least four jumps a year to maintain certification.

When it comes to air transportation, SEALs usually hitch a ride with the Air Force's Special Operations Wing or the Army's 160th SOAR ("Night Stalkers"), although the Navy does now have two helicopter squadrons that support special operations: the HSC-84 Red Wolves and HSC-85 Firehawks. These squadrons are accustomed to working around naval vessels, maritime assets (oil rigs, piers), rugged coastlines, and stormy seas. They fly the older but more combat oriented HH-60H Seahawk helicopter, and receive their highly specialized flight training from the Night Stalkers.

The SEALs have access to a number of state-of-the-art, heavily armed platforms designed for clandestine missions, including the CV-22B Osprey, MC-130H Combat Talon, MC-130J Commando II, MH-47 Chinook, MH-6M Little Bird, and MH-60 K/L/M helicopters. Of these, the helicopter is the most versatile. It can fly at high speed just feet above the ground, following the contours of the earth to thwart detection, and quickly insert/extract SEALs where the enemy least expects it using a variety of methods, including rappel, fast-rope, helo-cast, SPIE rig, or Jacob's Ladder. Helicopters can also conduct combat search-and-rescue, as well as serve as sniper and observer platforms.

In May 2011 it became known America had a secret helicopter design when DEVGRU and SOF/CIA operators took-down terrorist Usama bin Laden in Pakistan. The helicopter crashed during the mission and a tail section survived demolition, revealing its existence. It is referred to as the "Stealth Hawk" because of its alleged "whisper" flying mode and its use of radar-absorbent paint and composite materials that make it difficult for radar to detect.

Supporting the SEALs on missions are unmanned aerial vehicles. Ranging in size from 12 inches to 16 feet, they are used for communications relay, as well as scouting and battle damage assessment in real-time using video cameras that can see day or night. This provides SEALs (and higher-level commands) with better situation awareness. Some drones have a laser-targeting device so the SEALs can pinpoint a target for a missile strike, while others are armed with small rockets for immediate attack.

Rather than burden SEALs with flying and maintaining these drones, NSW sends along an Unmanned Aerial System Troop detachment. The men, who have undergone training to accompany SEALs on missions, are experts in piloting, surveillance techniques, and imagery acquisition. Drones used by the SEALs include (by size): the hand-held *Wasp* and *Raven*, the *Aqua Puma* (which can land in water), the *ScanEagle* (which has a range of 900 miles), and the brand new *Blackjack*. The latter can fly at 19,000 ft. and conduct surveillance for 16 hours, automatically identifying targets (tanks, vehicles, missiles). ★

The Air Force / SOCOM variant of the Osprey (CV-22) is used for long-range insertion, extraction and resupply of SEALs and other elite teams. It can transport up to 18 combat-ready men 620 miles and return, or fly 2,400 miles on one refueling. The craft's 38-foot diameter, gimbaled rotors allow it to fly like a turboprop airplane and hover like a helicopter, giving SEALs plenty of flexibility. The transition from helicopter to plane mode takes only 12 seconds. The Osprey, which has a cruise speed of 320 mph, augments the Combat Talon (MC-130) aircraft and features terrain-following/terrain-avoidance radar, FLIR, and an advanced electronic-warfare suite. All together, they enable the CV-22 to operate at low altitude in high-threat environments.

Above – Specialized naval units that support the SEALs in combat, such as SWCC and EOD (shown here), are all parachute qualified so they can deploy with them. Bomb technicians detect and render safe explosive devices used by enemy combatants, as well as help SEALs set booby-traps and demolition charges. This photo shows them practicing a static-line jump from a Hercules (C-130) at an altitude of 1,000 feet.

Right – High altitude jumps (13,000 - 45,000 feet) require parachutists to wear thermal clothing (it can reach -60 degrees at altitude) and use oxygen. The oxygen mask system shown here has the oxygen hose affixed to the bottom left side of the regulator to ensure it doesn't block the parachutist's vision. (Being able to navigate, read the altimeter, and find and pull the parachute ripcord is critical.) The hose goes over the left shoulder, runs under the parachute, and connects with oxygen bottles stored in a pouch on the right hip. Oxygen is supplied only when the parachutist inhales rather than flowing constantly into the mask.

Right – SEALs from Naval Special Warfare Unit 3 and EOD technicians leap from a Seahawk helicopter during a free-fall jump training exercise in Bahrain. The SH-60 can also be configured to support static line parachute jumps. A rectangular-shaped, wire anchor line is bolted to the cabin floor allowing up to eight SEALs to snap on to it and jump – four from each side of the helicopter. The edges of the cargo doors are padded and taped to prevent parachutists' gear from getting snagged and their static lines from being cut or frayed. When SEALs jump, the Seahawk must slow to 70 knots and fly no lower than 1,500 feet above the ground. To do otherwise would result in a parachute not opening fully, or even collapsing.

Below – Junior members of SBT-20 practice basic free-fall positions in the vertical wind tunnel at Fort Bragg before attending a more advanced military parachuting class. This student is in the Stable Free-Fall position, which will keep him from tumbling and spiraling while falling at 75-120 mph. To execute a left or right turn, the parachutist arches his back and looks in the direction of the turn while rotating his shoulder in the same direction. Before reaching the new desired heading he counters, thereby stopping the rotation. Then he assumes the Stable Free-Fall position again.

Above – Naval Special Warfare has a detachment at Key West, FL for advanced, year-round parachute and maritime-operations training. This photo shows a SWCC pulling his ripcord after free-falling from 12,500 feet. The SEALs are presently increasing their capability to perform high-altitude jumps up to 45,000 feet. From that height they can glide 30% further to a target than they can from 35,000 feet. The drawback is greater risk of hypothermia, hypoxia and decompression sickness, as well as longer prebreathing times (over an hour) at 10,000 feet to purge their bodies of nitrogen before ascending higher.

Right – Static line jumps, in which the parachute automatically opens as each man exits, are done at low altitude (1,000 - 3,500 ft.) to quickly put SOF personnel on the ground and avoid enemy fire. The men walk off the plane ramp at a 30-degree angle (left or right as ordered) so they are not hit by static lines and parachute bags. The jumpmaster spaces the parachutists apart so they don't collide in the air.

Right — The Leap Frogs Parachute Team, based out of Coronado, CA, performs precision aerial maneuvers at events nationwide to heighten public awareness about the SEALs, as well as assist with the Navy's overall recruiting efforts. The team is comprised of about 10 enlisted members, including SEAL and SWCC operators, a parachute rigger, a safety officer, a medic, and a public affairs specialist. It is led by an officer, who is usually a SEAL. Members serve on the team for three years and then return to their Naval Special Warfare unit. The Leap Frogs' publicity slogan is "Heroes on the ground. Legends in the sky." When not performing, the team assists in training SEALs in air operations.

Below — SEALs and German military forces free-fall parachute onto a frozen lake in northern Norway during a cold-weather training exercise. Although SEALs are broadly trained to wage war anywhere, each team also specializes in different environments and/or geographical region. For instance, SEAL Team 1 is the subject-matter expert on the Western Pacific and jungle/desert warfare, while SEAL Team 5 focuses on Korea and the arctic environment. This means Naval Special Warfare always has a highly trained unit that can respond with the correct equipment and proper combat tactics for a given area of the world.

Left – *The military recently replaced the MC-1 static-line parachute (shown here) with the MC-6 parachute, which is more controllable (it can make a 360-degree turn in just four seconds). The chute also has a slower rate of decent (15 fps), and can support greater weight (400 lbs. versus 360 lbs.). When the parachutist is about 250-feet above the ground, he begins lowering his equipment bag on a 15-foot-long line so it lands first, making his own landing smooth, safe and under control. SEALs jump at least quarterly to maintain their parachute certification(s).*

Right and Bottom – *Special Boat Teams are able to insert a 20,600 lb. RIB into any body of water using a C-130, C-5 or C-17 transport flying at 1,500 to 3,500 feet above the surface. Known as the Maritime Craft Aerial Delivery System (MCADS), a drogue pulls the RIB and its pallet out of the aircraft, where they then separate mid-air and fall. Both the RIB and pallet have their own parachutes, which deploy. The RIB subsequently lands in the water and is ready for use. (The pallet is allowed to sink in actual combat.) The SBT members quickly free-fall parachute after the RIB and land as close as they can to avoid wasting time. They are trained to have the boat ready and running within 15 minutes. SWCCs involved with MCADS undergo "wet silk" water jumps and egress training every six months to keep their skills honed.*

Above – *A SEAL assault team fast-ropes onto a ship and sprints into position to rescue hostages and take control of the vessel. The helicopter in the background with its cargo doors wide open is one of two sniper helicopters. (The other is off-page to the left.) The SEAL sniper, who is strapped in, shoots armed threats ahead of the assault team, warns the men of booby traps and ambushes, and helps guide the SEALs to the ship's bridge. The helicopter is intentionally at a 45- to 60-degree angle to the ship so the sniper's bullets do not skip into the SEALs on deck.*

Right and Next Page – *A SEAL fireteam fast-ropes from a SH-60 to the edge of a cliff during training. Fast-roping in the US military is restricted to special operations units. The rope measures 1.75-inches in diameter and comes in lengths of 50-, 60-, 90- and 120-feet. It's capped at one end with a 3-inch ring for aircraft anchoring. SEALs carry their assault rifle muzzle down at their side when fast-roping (opposite) so it is readily available for use once on the ground.*

Top – Assault dogs do everything SEALs do, including parachuting, helo-casting and rappelling. This photo shows a "Hair Missile" being lowered by hoist from a helicopter flown by the "Red Wolves" (HSC-84), one of two units that support SEAL and SWCC missions.

Right – Both the RIB and SOC-R can be extracted and transported by helicopter, giving Special Boat Teams greater range inland and more flexibility (e.g., avoiding river obstacles). It's a dangerous operation since the helo hovers a few feet above the boat during hook-up, and the men must ascend a ladder to climb aboard the helicopter via an underside "Hell Hole."

Left – Members of SDVT-1 climb aboard an Army Black Hawk during helo-cast training at Kaneohe Bay in Oahu. In 2008, Naval Special Warfare began consolidating its undersea assets in Hawaii. All SDV Teams were combined under SDVT-1 and headquartered out of Building 987 at their Pearl City compound. A 40,000 sf "Waterfront Operations Facility" was subsequently built to provide the platoons with training, planning and mission support space. It will be joined in the near future by an 86,000 sf "Undersea Operational Training Facility."

Right – A Seahawk flown by the "Firehawks" (HSC-85), a Navy helicopter squadron that supports SEAL missions, swoops in to extract pilots and air crewmen from a simulated downed aircraft during a nighttime escape-and-evade exercise in the desert.

Below – A CH-46 Sea Knight extracts members of SEAL Team 5 from the ocean. The SEALs spread themselves out in-line at 50-meter intervals in the water. A "Jacob's Ladder" is suspended from the rear ramp and trailed behind the helicopter. As it passes each man, he hooks an arm into a rung and climbs up, using his upper-body strength to fight the wake tugging at his legs and waist. In this manner, the SEALs are quickly retrieved one-after-another while minimizing the helicopter's exposure to enemy fire.

A SEAL sniper takes a peek through the sights of the accurate Mk-11 7.62mm, semi-automatic rifle (with sound suppressor). This is the same sniper rifle used by SEALs to take-out pirates holding cargo ship captain Richard Phillips hostage aboard a lifeboat in 2009. It is being replaced by the Mk-20 Sniper Support Rifle.

Above and Right – Although fast-roping is the preferred method to get SEALs quickly aboard a ship without having to land a helicopter, it's risky since the rope is attached to the men only by their hands and feet. That's why they do not carry heavy loads. The SEALs control their descent by the grip of their gloved hands and boots, and brake about two-thirds the way down to avoid landing on the man ahead of them or injuring themselves. About 10-feet of rope is kept on the ground to help anchor the loose end so it doesn't flail around in the rotorwash. Once aboard the ship, the SEALs organize into 4- or 6-man "trains" (right) and speed toward their assigned targets. The priorities are saving hostages and taking control of the ship. As for the MH-47 Chinook helicopter shown here, it's flown by the Army's "Night Stalkers" and is specially designed and equipped for special operations missions. The highly skilled pilots can even dip the Chinook's rear ramp several feet into the ocean to allow a SEAL fireteam to drive their speeding Zodiac directly up and into the back of the helicopter.

Right – A squad from a SEAL Team disembarks from an Army CH-47D Chinook helicopter during training at the Joint Pacific Alaska Range at Fort Wainwright, AK. The helicopter is practicing two-wheel landings on the rugged terrain while keeping its front wheels airborne. The SEALs spread out in a predetermined manner to establish control of the landing zone and engage targets, while minimizing their exposure to potential hostile weapons fire. JPAR is the largest training venue in Alaska, comprising nearly 2,500 square miles of land, 42,000 square nautical miles of ocean, and 65,000 square miles of airspace. Military forces can be fully integrated in realistic, large-scale, live-fire maneuvers. SEALs train there to hone their combat skills for mountainous and cold weather environments, such as Afghanistan.

Below – SEALs establish a perimeter around a helicopter during a combat search-and-rescue (CSAR) training mission for a downed pilot at NAS Fallon in Nevada. The men crouch low not only to minimize their exposure to enemy fire but also to avoid the spinning rotor blades, which are angled downward at the front end of the craft. This is why helicopters are always boarded from either side and never from the front. CSAR is not the only training offered at Fallon: NSWG-2 hosts a 19-day tactical course on how to drive the SEALs' ground mobility vehicle in combat.

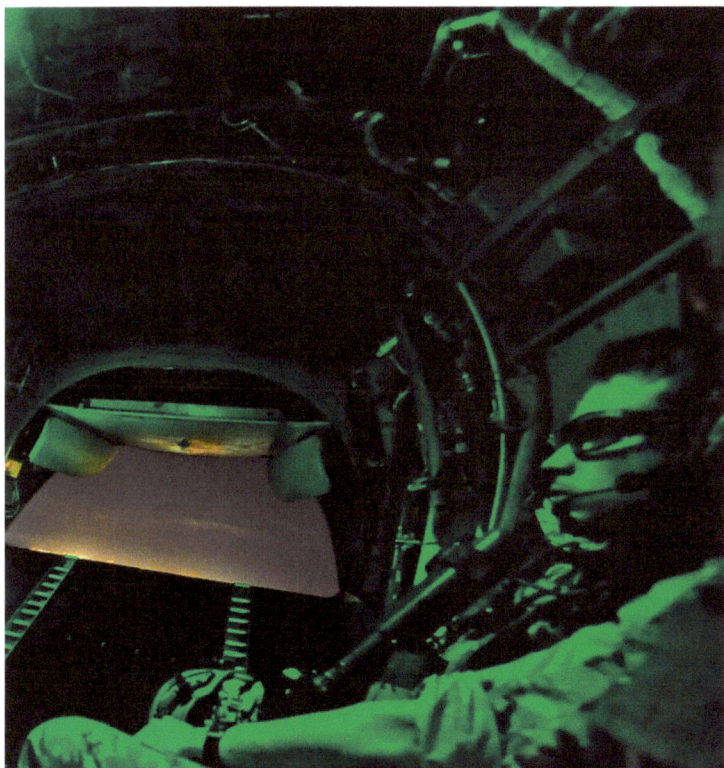

Above – A helicopter provides cover for SEALs being transported in Mine Resistant, Ambush Protected vehicles. The armored MRAP is used for high-risk missions that will likely involve enemy fire, mines and roadside bombs. The helicopter gunner warns the convoy of obstacles, ambushes and enemy forces.

Left – A night vision view inside the CV-22 Osprey.

Opposite Page – Special operations personnel wear heavy leather gloves (even welders gloves) when fast-roping due to the friction created when sliding down the rope. The men grasp the rope with both hands and pinch the rope between the insoles of their boots, and form an L-shape. Only two men are allowed on the rope at any given time, spaced 10-feet apart. They have to be aware that the powerful rotor-wash from the helo or CV-22 Osprey (as shown here) will push them downward faster, so they must control their descent, as well as be prepared for a minor shock from static electricity when they touch the ground.

When helo-casting from an Army MH-47 Chinook, the ramp is positioned 10-degrees below horizontal to facilitate jumping. If a Zodiac is needed for the mission, it is partly deflated ("Soft Duck") and then folded and stuffed inside the helicopter's cargo bay. (The motor, fuel bladders and other gear are secured to the floor of the boat under cargo netting.) The SEALs push the Zodiac out first and then jump after it (taking care, of course, not to land inside). A foot pump is then used to fully inflate the boat.

ACROSS THE
LAND

SEALS never just pack-up and respond to a crisis. Rather, they are part of a 24-month rotational training and deployment cycle that hones their combat skills and brings them up to fighting speed. When they finally are "on-call," it's as part of a Naval Special Warfare Squadron attached to a war-fighting command. The Squadron is comprised of SEALs – organized into four Task Units with two platoons each – and an extensive mission-support team.

The deployment cycle is broken into four, six-month phases. The first phase, Professional Development, provides each SEAL with the opportunity to receive advanced training in his specialty. Examples include foreign language, tactical driving, dive supervisor, special reconnaissance scout, tactical surveillance operations, helicopter rope suspension training, etc. These courses are offered by the NSW Advanced Training Command (Imperial Beach, CA), which oversees 84 courses and 14 special training sites.

One of the newest schools is the NSW Explosive Center of Excellence at Fort AP Hill. It's the military's preeminent breacher training facility. In addition to classrooms, it features an area containing a five-story breach house, door and wall mock-ups, a ship facade, a train and railway, etc. that allow SEALs to practice breaching techniques using ammunition, explosives, torches, and cutting saws. They also learn how to use fiber-optic systems or a "Throwbot" (a soda-can sized robot equipped with a video camera) to conduct recon in support of a particular breaching strategy.

The second phase, Unit-Level Training, is where the SEALs come together to refresh their combat skills while organized as a platoon, squad and fireteam. The specific type of training they receive depends on where they will be deployed and what type of missions they will likely be assigned. However, everyone undergoes training in weapons, land navigation, close-quarters combat, reconnaissance, night operations, patrolling, air operations, etc.

A critical skill is mastering how to respond to sudden enemy attack. That's because SEALs typically operate in small numbers on short-duration missions (e.g., 36 hours

Land Warfare

Overleaf – SEALs practice navigating rugged terrain in a Ground Mobility Vehicle during training at the NSWG-1 Desert Warfare Training Facility at Camp Billy Machen near Niland, CA. The GMV is the special operations version of the "Humvee." It's being replaced by the "Flyer," a lightweight tactical vehicle.

Left – A SEAL fireteam moves effortlessly through the sandy terrain during a live-fire, combat exercise. Take note of the control each man has over his weapon while running. Likewise, how the unit keeps an eye on their flanks for the presence of hostile forces.

for a desert foot patrol) and, hence, are not equipped for prolonged gun battles. They fight light. Attacks are survived through cunning, speed, and violence of action. In general, this means seeking cover and responding with 10-15 seconds of intense, accurate weapons fire (and grenades) to stun the enemy, and then "peeling off" to break contact. SEALs practice this during Immediate Action Drills.

Unit-level training is overseen by ATC Training Detachments on the east and west coasts. Some of the SEALs' specialized training facilities include: the Desert Warfare Training Facility at Camp Billy Machen, the Mountain Warfare Training Facility and the Assault Training Center of Excellence at Camp Michael Monsoor, and the Remote Training Site at Warner Springs. The latter is used for clandestine reconnaissance training. Four to six SEALs are inserted into the mountains by helicopter. They then spend three days conducting surveillance on designated targets while avoiding enemy forces, before finally being extracted.

Phase three, Squadron Integration Training, involves all components of the soon-to-be deployed NSW Squadron. Participating units varies with the deployment, but it can include: SEALs, EOD, SWCCs, SDVT, UAS Troop, Combat Support, Combat Services Support, a Communications Detachment, and a Canine Detachment.

Their combat skills now refreshed, the units jointly engage in large-scale, live-fire, scenario-based field exercises. Due to the number of combatants, the nature of the weapon systems used, and the full-scale size facilities needed, these maneuvers are held at large military reservations. The Navy's San Clemente and San Nicholas islands lend themselves to maritime and over-the-beach scenarios.

The Army's Camp Roberts is ideal for land warfare involving artillery. And the Marines' Twentynine Palm's 300-acre, 1,500 building mock city is perfect for urban, desert-warfare scenarios.

One facility the SEALs often use is the Joint Maneuver Training Center at Camp Atterbury. It features an austere training environment in a secluded and secured area where they can immerse themselves in realistic combat conditions 24-hours a day. It also offers a 1,000-acre urban setting that includes a large city (including houses, school, hospital, churches, refinery, prison, marketplace, stadium, and embassy), nine miles of roads, collapsed structures, and a searchable tunnel system. Role players who are fluent in different languages and cultures are used to populate the city, and exercise controllers can create sound, smell and smoke as needed to make the training as realistic as possible.

Squadron Integration Training is controlled by the Naval Special Warfare commodores on each coast of the United States. Depending on where the Squadron is ultimately headed, it could also benefit from environmentally specific training in, for example, cold weather, mountain warfare, and jungle and riverine warfare. So the Squadron could find itself in the chilly tundra of the Joint Pacific Alaska Range Complex or in the mosquito-ridden swamps of the Western Maneuver Area at the Stennis Space Center.

The final phase is Deployment. For a half-year the NSW Squadron is assigned to a regional combatant commander (or theater special operations command, or a naval fleet) to respond to contingencies. SEALs are experts at reconnoitering and conducting covert attacks, and they often find themselves deploying on short notice and in small teams to handle a situation. Typical missions include: tagging, tracking or eliminating high-value "items of interest;" providing real-time, first-hand intelligence for military decision makers; and secretly emplacing surveillance sensors on land and in the ocean to monitor the enemy. The latter includes SOCOM's classified worldwide intelligence-gathering network. Aerial drones and remote ground sensors (some of which resemble rocks) collect seismic, acoustic, magnetic, infrared and full-motion video data and transmit it to a Mission Support Center for analysis and tasking. The Navy's maritime component of this spy network (*formerly called "Sea Eagle"*) collects and shares intel provided by undersea sensors, SDVs and UUVs. ★

A SEAL recon team quietly patrols the forest during a mountain warfare training exercise. Both are armed with the Mk-17 7.62mm semi-automatic rifle, which has earned high marks for its accuracy and stopping power. Switchable barrels allow it to be changed from a combat assault rifle into a CQC weapon or quasi sniper rifle (with an 800-meter range). If needed, the Mk-13 grenade launcher can be affixed below the barrel and to the ammo magazine. It fires 40mm grenades out to 600 meters at a rate of five rounds a minute. Take note of the sound suppressor affixed to the rifle held by the rear SEAL.

Above – SEALs wade ashore after being dropped into the icy coastal waters off Kodiak Island in Alaska as part of a cold-weather training course. A detachment from the NSW Advance Training Command is based at Spruce Point on Kodiak (130-acres leased from the US Coast Guard through 2021). Using nearby Long Island and the vast tracts of uninhabited land on Kodiak, the instructors teach SEALs how to survive, navigate and wage combat in this tough climate, as well as understand how their weapons and gear will function. SEALs also train at the Army's Northern Warfare Training Center at Black Rapids, Alaska.

Right – A SEAL watches as munitions discovered during a search of an Afghan village are destroyed. SEALs are trained to gather intelligence (documents, computer data) when conducting a sensitive-site exploitation mission, as well as to collect measurable biometric data (fingerprints, iris scan, facial features) of local leaders and suspect combatants using portable devices like the Cross Match SEEK II.

Right – Naval Special Warfare's Western Maneuver Area at the Stennis Space Center is used not only for Special Boat Team riverine training but also by SEAL Teams to conduct live-fire exercises, such as the "Immediate Action Drill" shown here. An IAD ensures the men know how to instinctively respond to a sudden enemy attack and escape from the kill zone.

Below – The M60 machine gun is a SEAL favorite since it delivers tremendous firepower with minimal recoil. The average SEAL shoots more than 100,000 bullets a year in training. Such live-fire facilities include the 42,784 acre Camp Roberts (shown here) located north of Paso Robles, CA. It offers 25 training areas, two airfields (including a facility and airstrip for unmanned drones like the RQ-7 Shadow, RQ-11 Raven, RQ-20 Puma, and RQ-21 Blackjack), parachute drop zones, weapon ranges, rappelling and mountaineering sites, NBC chambers, helicopter nape-of-the-earth flying zones, and a mock city for urban combat. The latter is comprised of 14 buildings (including a town hall, hotel and church) and features an underground tunnel complex, as well as an audio and smoke system that simulates realistic conditions. So if a training scenario requires a call to prayer from a mosque, the smell of an open-air market, or machine-gun fire and the smell of rotting corpses, the facility can provide it.

Above – The load-bearing assault vest puts everything a SEAL needs during combat easily at hand. Extra 30-round magazines for the M4A1 are carried in the three front pockets, allowing for a fast reload. (Note that the SEAL shown here keeps his weapon out of the dirt so it doesn't jam.) The administrative pouch on the chest holds maps, etc. as well as an ammo clip for the sidearm. The Velcro area is used to display unit and other identification patches. Behind the pouch is a slot that holds a ballistic chest plate to protect the SEAL from small arms fire and fragments.

Right – The cold, moist woodlands of coastal Alaska provide a challenging environment in which to train, especially during the winter months. SEALs fight off the constant threat of hypothermia, which can cause shivering, confusion, clumsiness, stumbling, slurred speech, drowsiness, and poor decision making. All of these impact a SEAL's ability to maneuver, fire weapons and maintain situational awareness, thereby jeopardizing the success of the mission.

Right – Every SEAL knows that casualties are an inevitable consequence of war regardless of what is done to limit the risk. They consciously accept that reality, understanding they might indeed be the next man to be dragged out of a firefight by his vest. The SEALs shown here simulate evacuating a wounded teammate during an Immediate Action Drill at the Western Maneuver Area. No SEAL is <u>ever</u> left behind on the field of battle.

Below – Unlike other military dogs that are trained for just one specific role, SEAL combat-assault dogs ("Hair Missiles") are cross-trained in attack, tracking and explosives. German Shepherds, Dutch Shepherds and Belgium Malinois are the breeds handpicked by their SEAL handlers from European breeders. Each dog team attends an eight-week basic course, followed by three months of field training. A deploying SEAL squadron is assigned two dog teams. The dogs wear ballistic tactical vests to protect them from injury. A lightweight video camera can be affixed to the vest so it can transmit real-time imagery to the SEALs. For night operations, the dogs are outfitted with "doggles," which offer night vision and infrared capability. They are also equipped with earpieces so they can receive commands. The Belgian Malinois in this photo is named Astor. Like all combat dogs, he is considered to be a full member of the United States military.

Left – One never knows when an enemy will resort to using nuclear, biological, or chemical weapons. Because of this, SEALs train to operate in such lethal, contaminated environments. Although donning protective clothing and a gas mask will protect them from most threats (it does not protect from gamma radiation), it is hot, claustrophobic, and physically and mentally taxing. Likewise it muffles communication, limits visibility, and restricts ease of movement. Yet SEALs persevere so they can accomplish their mission.

Right – Arctic, mountain, and extreme cold weather environments are challenging for SEAL operations. SEALs must know how to evaluate the weather and terrain, plan over-snow movements, use snowshoes, haul an ahkio sled, and maintain weapons and other gear in the cold. To keep warm and alert, they eat MCW/LRP rations that provide 4,500 calories a day.

Below – Just because you're down doesn't mean you can't fight. SEALs learn many combat tactics and techniques to overwhelm hostile forces. For instance, low grazing fire is aimed at the feet of enemy soldiers. It is not only effective at wounding directly and through ricochets (soldiers tend to hit the ground when the shooting starts), but the dirt and debris thrown up by the striking bullets adversely impacts the enemy's ability to see, shoot and exercise control.

SWCCs take their motto, "On Time. On target. Never quit!" very seriously, understanding that if they are even a minute late arriving where they need to be, the mission is jeopardized and SEALs can die. To help them meet their obligations, Naval Special Warfare is buying faster and better equipped combatant boats, developing a helmet-mounted, heads-up display system that provides situation awareness, and acquiring an additional 1,640 acres of land along the Pearl River at the Western Maneuver Area for them to conduct live-fire .50 caliber machine-gun training.

Above – Naval Special Warfare built a 26-building mock city at the northwest corner of The Rock (shown here) so SEALs can practice live-fire urban warfare tactics. The city includes a five-story hotel, city hall, church, bank, embassy, school, and a bazaar-like market. The narrow streets and winding corridors are modeled after cities found in developing countries. Here, SEALs learn how to conduct urban patrols and respond to riots, ambushes, snipers, booby traps, and hostage situations. The concrete-block buildings are designed to withstand flashbangs and breaching charges detonated inside.

Right – SEALs practice room-clearing tactics in a kill house located at the 7,000-acre Academi facility in Moyock, NC. The walls are movable, allowing for different configurations. Paper targets representing hostages and armed terrorists ("tangos") are tacked in different locations. SEALs must identify and kill the tangos with a "double-tap" (two bullets fired in succession) to the head or chest as they rush into the room.

Right – A female hostage is led away by a SEAL after being rescued during a simulated hostage-rescue mission. SEALs rely on Surprise, Speed, and Violence of Action when conducting such operations. First, they gather intelligence about the facility, hostages, and terrorists so a plan can be devised and rehearsed. Then they attack unexpectedly from multiple indirect approaches (often at night) and use distractants (flashbangs, smoke, CS gas) and precise firepower to overwhelm the terrorists before they can respond. Their goal is to secure the hostages within 30 seconds. Although all SEAL Teams are trained for hostage rescue, DEVGRU is the Navy's premier unit for counterterrorism and hostage-rescue missions abroad. Domestically, that responsibility goes to the FBI's Hostage Rescue Team.

Below – Since EOD technicians accompany SEALs into combat to handle booby-traps and improvised explosive devices, they must be skilled at many of the same tactics used by the SEALs, including building entries (as shown here). While the EOD tech on the far right provides security, the remaining four-man "train" will enter the building with the first two men criss-crossing as they pass through the door. Each man in the train has an area of the room he is responsible for clearing. As a team, they will go room-to-room until the entire building is under their control.

Left – A Sniper-Observer (SO) team hides in the open shadows of a grove. The vegetation in this photo has been cut locally and affixed to the SEALs' ghillie suits, transforming them into shrubs. The sniper (right) is armed with the powerful Mk-15 .50 caliber rifle, which can penetrate truck engines and some armored vehicles, knockout generators and communication arrays, and kill combatants hiding behind walls. The observer (left) is using his rifle scope to determine target selection and tracking, range, wind strength and direction, etc. DARPA, the Pentagon's research arm, has developed a steerable .50 cal. bullet ("EXACTO") that can make its own adjustments mid-flight so that it hits the intended target, even if it suddenly moves. It will make snipers even more lethal.

Right – A SEAL sniper deployed in Rawah, Iraq takes aim with the lightweight 5.56mm Mk-12 rifle. The Keffiyeh headdress camouflages his face, shields him from sunburn, and protects his eyes from blowing sand. Most SEALs call it Shemagh (pronounced "schmog") rather than Keffiyeh.

Below – This photo shows some of the technology used by SO teams to acquire and monitor their targets, as well as send reconnaissance data to headquarters via a secure satellite link. The sniper, with his Mk-15 rifle, is laying prone behind the observer (front), blending perfectly with the vegetation and terrain features.

Left – The Lightweight Tactical ATV offers SEALs speed and agility in all environments. It carries two SEALs and 500 lbs. of gear at speeds up to 48 mph. An optional swing-arm mount on the passenger side holds a M240 machine gun. SEALs gain hands-on experience driving the LTATV to its limits during an intense, five-day course taught by Overland Experts. The class teaches terrain analysis and navigation, obstacle avoidance, extraction of stuck vehicles, field repairs, and driving while wearing NVGs. One of the training locations, the rugged Ocotillo Wells State Vehicular Recreational Area (shown here), prepares SEALs for operations in countries like Afghanistan.

Right – A SEAL sprints across the beach carrying the 7.62mm M240B machine gun. The belt-linked ammo is fed directly from the semi-rigid container.

Below – All SEALs learn to drive five military vehicles, including the GMV (shown here). A more advanced, 19-day "Tactical Ground Mobility" unit level class for SEALs is held at NAS Fallon. There, the men learn how to fire and maneuver against enemy forces, employ "Figure 8" shooting tactics, drive in silent convoys, and fight in urban settings. Some GMVs are equipped with a long roof-mounted ramp. Using a hydraulic-lift system, the ramp is elevated so SEALs can breach aircraft and the upper floors of buildings.

Above – *The Mine Resistant, Ambush Protected vehicle (far right) is part of a family of vehicles designed to protect military forces from mines, IEDs and weapons fire. They feature a V-shaped hull (to deflect blasts), raised chassis, armored plating, all-terrain suspension, run-flat tires, a communications suite, and a gunner's turret. There are four MRAP categories: (I) for urban-combat missions (transports six); (II) for troop transport, convoy escort, and ambulance missions; (III) for mine clearing; (IV) the MRAP All-Terrain Vehicle, a smaller, lightweight combat model.*

Right – *Trained NSW women deploy with SEALs as part of a Cultural Support Team to help with language translation, tactical interrogation, and intelligence gathering. The women undergo a half-year qualification program that includes learning irregular warfare theory, mission planning, and combat skills (e.g., navigation, fast-rope, rappelling, convoy fundamentals, small unit tactics, first aid), as well as three weeks of weapons training with the M4 and M9.*

Right – People equate SEALs with machine guns, handguns and grenades, not mortars. However, the M224 60mm lightweight mortar can be hand held, as shown here during live-fire weapons training at Camp Roberts. The 18 lb. unit fires high explosive, smoke and illumination (visible light and infrared) rounds up to two miles at sustained rates of 20 shells per minute. As such, it's ideal for SEALs to use to take out a sniper position, hit a target hidden behind a hill (or cluster of buildings), destroy a bridge, scatter enemy forces, illuminate a target, etc.

Below – Using night vision, SEALs prepare to breach a building. Naval Special Warfare has many kill houses for SEALs to practice close-quarters combat, such as: a 26,500 sf East CQC Training Range (Little Creek, VA); an 80,000 sf Indoor Dynamic Shooting Facility at the Silver Strand Training Complex (Coronado, CA), and a 15,000 sf CQC Range at Camp Monsoor (La Posta, CA). Each has multiple rooms, bullet-absorbing walls and furniture, and breachable doors and walls. They can be configured as a home, urban setting, or even a ship. To rehearse a real-world mission though (e.g., when DEVGRU pursued bin Laden in Pakistan), SEALs build a full-scale model of the target in a remote, secure area like Eglin AFB, Fort Bragg, Camp Atterbury, a CIA facility, or even overseas at a base or forward-training area.

The Only Easy Day Was Yesterday...

Attack Board – A quasi clipboard containing a compass, watch and depth gauge. It is used for underwater navigation.
Banana – A BUD/S candidate.
Black Ops – Black operations. Clandestine, secret missions SEALs do not discuss.
Boat Guy – What SWCCs call themselves.
Boat Pool – On an insertion mission, when RIBs or Zodiacs stop and gather offshore. It's often done to deploy swimmer scouts.
Body Snatch – An operation to kidnap a high-value target.
Budweiser – The special warfare insignia worn by SEALs. So named because of its resemblance to the Anheuser-Busch logo.
Bullfrog – The SEAL operator with the most time in service, regardless of rank.
Coffin – Nickname for the Mk-8 SDV
CRRC – Combat Rubber Raiding Craft
CQC – Close Quarters Combat, special tactics used in buildings, ships, etc. in which space for movement, assault and retreat is limited.

Crash Back – A technique in which a combatant craft abruptly halts its forward motion within one boat length.
DA – Direct Action. A short-duration strike (and other small-scale offensive actions) to seize, destroy, capture, recover or inflict damage on personnel, facilities, etc.
DEVGRU – Development Group, the Navy's counter-terrorism unit based at Dam Neck, VA. It was formerly known as SEAL Team Six. Its members are known as "Jedi."
DOR – Drop On Request. When a SEAL candidate quits BUD/S. *See Mother Moy's Bell.*
Dry Deck Shelter – The cylindrical-shaped container affixed to a submarine's deck that hangars the SEAL's mini-submarine.
Eight Boat – Nickname for the Mk-8 SEAL Delivery Vehicle mini-submarine. *See SDV.*
Fast-Rope – A helicopter-insertion technique in which operators slide down a rope using gloved hands to control the rate of decent.
Flashbang – A small, non-lethal grenade that uses bright light (6-8 million candela), loud noise (140-180 decibels), and a blast wave to stun hostiles during a room entry.
FLIR – Forward-Looking Infrared

Glossary

A long-held tradition: When a SEAL dies or is killed in action, his teammates use their fists to pound their Naval Special Warfare Trident pin into the top of the coffin as a sign of respect. In this instance, an Army Special Forces operator has also added his patches.

Gray Man – A BUD/S candidate who doesn't attract the attention and wrath of the instructors. He blends in.
Grinder – The courtyard at the Naval Special Warfare Center where SEAL candidates perform a myriad of exercises for hours on end.
HAHO – High Altitude, High Opening
HALO – High Altitude, Low Opening
Hush Puppy – A suppressed handgun (often firing a subsonic round) used to kill guard dogs and sentries.
Kill House – A special building where SEALs practice dynamic entries and CQC skills using frangible bullets and live ammunition.
Laser-Guided Furball – What SEALs call their military-trained canines. Also referred to as "Hair Missiles."
Low Vis – Low visibility. Used in reference to clandestine operations.
Monster Mash – A grueling training session that consists of a 10-mile run, 2-mile swim, 3-mile boat paddle, and rock portage.
Mother Moy's Bell – A tugboat bell that is rung three times by BUD/S candidates when they want to quit ("Ring Out"). Named after Master Chief Terry Moy, an instructor at BUD/S, who gave it to Class 58.
NBC – Nuclear, Biological and Chemical
NOE – Nape Of the Earth. A flying technique in which aircraft fly only a few feet above trees, following the contours of the earth.
Non-Qual – A candidate who fails to graduate BUD/S.
NSW – Naval Special Warfare
NVG – Night Vision Goggles
Over-Whites – White-colored, nylon outerwear worn to camouflage SEALs in snow. Also called "skins."
Pool Rat – A SEAL trainee who is at home in the pool.
Pulk – A flat-bottomed sled used in snow.
The Rock – San Clemente Island
Room Broom – A compact submachine gun, such as the MP-5K. It "sweeps" a room with bullets in seconds.
Rusty Trusty – The fixed-blade knife carried by SEALs. It's often rusty due to saltwater exposure.
SBT – Special Boat Team. An integral part of naval special warfare operations. *See also SWCC.*
SDV – SEAL Delivery Vehicle, a mini-submarine used to transport SEALs underwater to shore. *See Dry Deck Shelter.*
SEAL – SEa, Air and Land. Also: "Sleep, Eat And Live it up."
Side-by-Side – Nickname for the SEALs' lightweight all-terrain vehicle.
Slick – When a SEAL forgoes body armor.
Sneak And Peak – A recon operation.
SOCOM – US Special Operations Command
SOC-R – Special Operations Craft, Riverine. Pronounced "soccer," this is an armed, high-speed, shallow-draft boat operated by SWCCs to insert and extract special operations forces.
SOF – Special Operations Force. Used in reference to any military unit involved in clandestine operations.
SR – Special Reconnaissance. To gather specific, time-sensitive intelligence of strategic or operational importance in hostile or politically sensitive areas.
Steel Pier – A steel barge that soaked BUD/S candidates are ordered to lay on for hours at a time. Their body temperature drops to near hypothermic levels.
SUBOPS – Submarine Operations
Sugar Cookie – A soaking-wet BUD/S candidate who is covered in sand, resembling a sugar-sprinkled cookie.
SUROBS – Surf Observations
SWCC – Special Warfare Combatant-craft Crewmen. SWCC (pronounced "swick") operate the boats used to insert and extract SEALs. Some SEALs teasingly define SWCC as: "SEAL Wannabe: Couldn't Cut it."
Tadpole – A BUD/S candidate.
Team Guy – What SEALs call themselves.
The Teams – The Naval Special Warfare community.
Tug Line – A cord that when abruptly tugged silently alerts resting SEALs on patrol of approaching danger. The cord is wrapped around each SEAL's finger.
Turtlebacking – Long distance underwater and surface swimming while wearing UBA.
Twin 80s – The two SCUBA air tanks worn by SEALs in BUD/S, each of which holds 80 cubic feet of oxygen.
UBA – Underwater Breathing Apparatus
UUV – Unmanned Underwater Vehicle
WARCOM – Naval Special Warfare Command

Acknowledgments

Sincere appreciation is extended to: Captain Dawn E. Cutler, USN, Deputy Chief of Naval Information; Senior Chief Petty Officer Joseph D. Kane, MCCS (SW), Public Affairs LCPO, Naval Special Warfare Command; MC1 Dominique Canales, Public Affairs LPO, Naval Special Warfare Command; MC1 Eli J. Medellin, formerly with Combat Camera Group Pacific; Scott Williams, Navy SEAL & SWCC Scout Team, Naval Special Warfare Command; LCDR Matthew R. Allen, Public Affairs Officer, SOCOM; Mike Bottoms, Managing Editor, *Tip of the Spear*, SOCOM; and, from a visit many years ago: the Naval Special Warfare Center, SEAL Teams Two and Five, and Special Boat Squadron One.

Unless noted otherwise, images in this book are courtesy of numerous combat photographers and mass communication specialists in the US Navy and throughout the US Department of Defense. From the metadata provided with each image (when available), photo credit is given to the following photographers: MC2 Dominique Lasco (cover, 18-19, 21 top); Cherie A. Thurlby (back cover top); Chief Photographer's Mate Andrew McKaskle (1, 25 top, 50, 64 bottom, 65, 66 both, 67 top, 68 both); Staff Sgt. Chris Griffin, USAF (6); MC2 John Scorza (9 top, 16, 103 top, 105 top, 120); MC2 Gary L. Johnson III (9 bottom); MC2 Kyle D. Gahlau (12, 14, 17 bottom, 21 bottom, 24, 28, 29, 80 bottom); Photographer's Mate 3rd Class John DeCoursey (15); MC2 Marcos T. Hernandez (17 top, 20 top); Mass Communication Specialist Seaman Michael Lindsey (20 bottom); MC2 Arcenio Gonzalez Jr. (22-23); MC2 Dominique M. Canales (25); MC2 Christopher Menzie (26-27); Seaman Stephen M. Fields (30); MC2 Michelle Kapica (31 bottom); MC3 Blake Midnight (32-33, 34, 37 both); MC2 Christopher Menzie (35, 36); Chief Mass Communication Specialist Kathryn Whittenberger (40, 42-43, 61 bottom, 83 top); Journalist 3rd Class Davis J. Anderson (44 top); MC3 Robyn Gerstenslager (45 top); Photographer's Mate 1st Class Robert McRill (45 center); MC2 Joshua T. Rodriguez (45 bottom); MC3 Adam Henderson (46-47, 49 top, 57, 58); MC2 Meranda Keller (52 top, 93 top, 116 top); Technical Sergeant Brian Snyder, USAF (52 bottom); Photographer's Mate 2nd Class Ted Banks (56 top); MC1 Jayme Pastoric (56 bottom); Photographers Mate First Class Brien Aho (59 bottom); Mass Communication Specialist Seaman Apprentice Brian Read Castillo (60, 88-89); Photographer's Mate 2nd Class Eric S. Logsdon (61 top, 98, 102 top); MC3 Alex Smedegard (62-63); Lt. Commander Wood, USN (67 bottom); Capt. Richard Barker, USA (70-71); MC2 Ashley Myers (72); Photographer's Mate 1st Class Steven Harbour (74); Senior Airman Andy M. Kin, USAF (76-77); Photographer's Mate 2nd Class Scott Taylor (78 top); MC3 Anna Kiner (78 bottom); MC2 Michael D. Blackwell II (79 top); MC2 Gary L. Johnson III (79 bottom); Shane Hollar, USN (80 top); MC2 Matt Daniels (81 top); MC2 Matt Daniels (81 bottom); PH2 Crystal Brooks (82); MC2 Joseph M. Clark (83 bottom); Photographer's Mate 1st Class Michael W. Pendergrass (84 top); Chief Mass Communication Specialist David Rush (85 top); MC3 Robyn Gerstenslager (85 bottom); Capt. Richard Barker, USA (86); Mass Communication Specialist Seaman Apprentice Conor Minto (87 top); Sgt. Aaron Rognstad, USA (90 both); Lance Cpl. Ryan Rholes, USMC (91 top); Sgt. Aaron D. Allmon II, USAF (91 bottom); Staff Sgt. Tony R. Ritter, USAF (92); Senior Airman Sam Goodman, USAF (93 bottom); Airman 1st Class Christopher Callaway, USAF (94-95); MC3 Antonio D. Ramos (96); MC2 Martin L. Carey (100-101); Photographer's Mate 1st Class Tim Turner (102 bottom); Photographer's Mate 1st Class Arlo K. Abrahamson (103 bottom); Chief Mass Communication Specialist Robert Fluegel (105 bottom); MC2 Eddie Harrison (110 both); Tech. Sgt. DeNoris A. Mickle, USAF (111 top); MC3 Randy Savarese (111 bottom); Petty Officer 2nd Class Eli J. Medellin (113 top); MC2 Megan Anuci (114); Staff Sgt. Amber K. Whittington, USAF (115 top); Photographer's Mate 2nd Class Brandon A. Teeples (117 top); MC2 William S. Parker (117 bottom); and Sgt. Daniel P. Shook, USA (118-119).

And last but not least, a heartfelt thank-you is offered to Diane Duff for her inspiration and her unwavering support of this new book series. ★

About the Author

The son of a career naval officer, and the former *Homeland Security and Emergency Response Coordinator* at the New Hampshire Department of Health & Human Services, SF Tomajczyk has written extensively over the years about military affairs, terrorism, homeland security and emergency preparedness. He is a graduate of the University of Michigan and the New York Institute of Photography, and is listed in *Who's Who in America* and *Contemporary Authors*.

A born-again Christian, Tomajczyk is professionally affiliated with the Author's Guild, Academy of American Poets, Society of Children's Book Writers & Illustrators, and the National Eagle Scout Association. He resides in the "Live Free or Die" state of New Hampshire. ★

Books by SF Tomajczyk

SEALs: Naval Special Warfare in Action
To Be a US Marine
Black Hawk
US Counterterrorist Forces
Modern US Navy Destroyers
Carrier Battle Group
Bomb Squads
101 Ways to Survive the Y2K Crisis
US Elite Counter-Terrorist Forces
Dictionary of the Modern United States Military
The Children's Writer's Marketplace
Eyes on the Gold

Website

www.Tomajczyk.com

www.ingramcontent.com/pod-product-compliance
Lightning Source LLC
Chambersburg PA
CBHW042138290426
44110CB00002B/49